Disaster and Climate Change Preparedness in Maui

This project was funded by the American Psychological Foundation Visionary Grant, with support from the Pacific Regional Integrated Sciences and Assessments (Pacific RISA) program and the East-West Center.

Kati Corlew, PhD, is project principal investigator for this project, "Relating the Psychological Recovery from Recent Disasters to Climate Change Risk Perception and Preparedness in Hawai'i and American Sāmoa." She is an assistant professor of Psychology at the University of Maine at Augusta. She can be reached at kate.corlew@maine.edu.

For a free electronic file, available for download, and to learn more about the Pacific RISA project, visit www.PacificRISA.org.

The handbook is also available at EastWestCenter.org/Publications.

For permissions requests contact EWCBooks@EastWestCenter.org.

Disaster and Climate Change Preparedness in Maui
ISBN 978-0-86638-256-4 (print) and 978-0-86638-257-1 (electronic)

Photograph sources in this publication:
Cover by Krista Jaspers
Pages 1, 23, and 24 by Krista Jaspers
Page 3 by Laura Brewington, PhD
Page 6 by Kati Corlew, PhD

Overview

Maui is home to some of the most beautiful ecosystems on earth. But if you've lived here long, you know that Maui (like everywhere else) is vulnerable to natural and man-made disasters. Vulnerabilities include such risks as drought, wildfire, heavy storms and flooding, mudslides, erosion, tsunami, and earthquakes, among others.

Some of these hazards will be exacerbated in the coming years by changes to the climate. For this reason, disaster and climate change preparedness can go hand in hand. Many disaster preparedness actions will make families, business, and communities better prepared for climate change as well.

The purpose of this project is to better understand Maui's relationship with natural hazards and to help Maui citizens and professionals prepare for future events. This project connected with Maui community members about their experiences with hazard events in three different ways:

1) An online survey
2) Interviews with community members and professionals
3) A preparedness workshop in Wailuku

This booklet includes information about natural hazards and vulnerabilities to disaster in Maui, stories from project participants about their experiences, and a guide to disaster and climate change preparedness.

Hazards versus disasters

What's the difference?

A **hazard** is a threat of an event (natural, man-made, technological) that could be harmful to people or the environment.

A **disaster** is a natural, man-made, or technological hazard event that results in physical damage, destruction, loss of life, etc.

> Living in a flood zone is a *hazard*.
> The Japanese tsunami was a *disaster*.

A disaster can sometimes set off **secondary disasters**, or even create a series of cascading disasters, like a domino effect. They can include natural and man-made hazards.

> In 2011, an *earthquake* of the coast of Tohoku, Japan caused a *tsunami* that *destabilized* a nuclear reactor...

The International Disaster Database (EM-DAT) uses the following criteria as a guideline for defining a disaster:

- 10 or more people reported killed and/or
- 100 or more people reported affected and/or
- Call for international assistance/declaration of a state of emergency

Recognizing hazards in Maui

Maui features many diverse landscapes all on one island and also has a rainy and a dry season. This variability means that Maui is home to a wide variety of weather-related and non-weather related hazards that interact with each other to create many types of secondary hazards in different places around the island.

Periods of **drought** can cause the foliage to dry out, making parts of Maui more vulnerable to **wildfire** and **fugitive dust**. Then, **heavy rains** can cause **flooding**, coupled with **mudslides** that can carry earth, vegetation, and other debris down the gulches into shoreline communities.

Maui also has non-weather-related natural hazards like **tsunami** and **earthquakes**.

Natural hazards
are hazards to communities.

- Natural hazards can damage infrastructure, like roads and bridges, or electricity and water supplies
- Natural hazards can interrupt community functions – from family functions to work and school to shipping and societal organization
- Natural hazards can cause injury and loss of life

Climate change increases hazards and disasters

Climate change is already affecting Hawai'i, the Pacific Islands region, and the world. Though the severity and timing of the changes cannot be precisely predicted, scientists from the Pacific and around the world know many changes will happen.

In Maui, **many species are at risk**. Endangered species, already very vulnerable, are at increased risk:
- Silversword
- Honeycreepers
- Other spiritually and culturally important species

In Maui, **weather patterns** will become more extreme:
- Increased drought AND increased storm severity
- Flood, erosion, wildfire, runoff and mudslides

In Maui, **ocean acidification**, **erosion**, and other threats may damage the **coast and near-shore environments**:
- Unstable coastlines
- Lack of storm buffer

Remember, risks to the environment ARE risks to the community:

They threaten livelihoods, food security, and infrastructure.

Emergency Management Cycle

Disaster and emergency managers think about disasters as occurring in a cycle of stages, called the Emergency Management Cycle. Before the event is a time for **preparedness**. Immediately after the disaster strikes is the **response**. For the weeks, months, or even years after a disaster is the period of **recovery**. As the community becomes stable again, **mitigation** activities increase the community's ability to respond and recover from future disasters. This leads once again into **preparedness**.

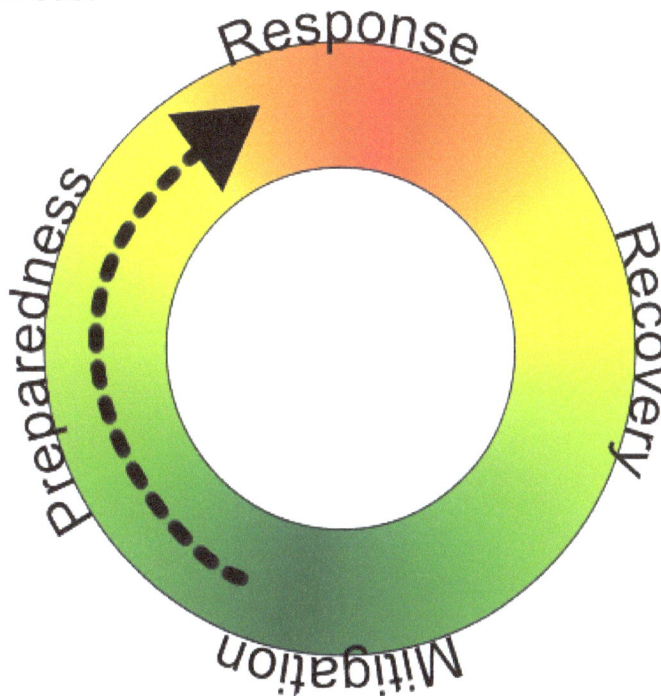

Image Source: http://en.wikipedia.org/wiki/File:Em_cycle.png

Experiencing Disaster
What happens psychologically before, during, and after a disaster occurs?

Many psychological experiences occur throughout the emergency management cycle: before, during, and after a disaster event. Some of these experiences are negative. Some are positive. Some are surprising, and though they are common, many people do not expect them.

Before the disaster event:

- Self-efficacy, or readiness – this can be increased by spending time preparing a disaster kit, making a plan, and learning about hazards
- Validating – when people know disasters are coming, they will often 'check in' with their loved ones to see if everyone else is making the same decisions they are, e.g., Are you boarding up your house? Are you evacuating?

During the disaster event:

- Proactive – when a disaster is occurring, people will take what time and opportunities they can to mitigate the impacts of the disaster, e.g., Grab your wallet or your shoes before evacuating from an unsafe location; close windows in a fire
- Reactive – when a disaster is occurring, people will often find themselves responding to events without conscious thought or reasoning; when events are happening quickly, there may simply be no time

Immediately after the disaster event:

- In the hours and days following a disaster event, people may experience anxiety, psychological trauma, and increased sense of risk; they will often find themselves expecting another disaster at any moment
- In the hours and days following a disaster event, people will often exhibit higher rates of helping others; mutual care for family, friends, and strangers; it is also common to see family and community mobilization as people help each other to recover from the disaster event

Ongoing recovery:

- In the weeks, months, and even years following a disaster, people may experience stress, sensitivity, intense feelings, and an on-going increased sense of risk to future disasters
- In the weeks, months, and even years following a disaster, people may also experience increases in self-efficacy, adaptation, and preparedness for future disasters. This is called **post-traumatic growth** and is associated with people learning what they are capable of doing during a time of extreme stress.

Heroic actions may occur at any point in this cycle. They may be proactive or reactive. They may be momentary or they may last for hours, days, or even years.

Survey results

What did people have to say about their disaster experiences? How prepared were they for future disasters?

The survey was conducted in Maui and American Sāmoa. Answers from 33 people were included in the final analysis – 14 (42%) from Maui, and 19 (58%) from American Sāmoa. Participants had both a personal and a professional interest in disasters. 17 (52%) were women, and 15 (46%) were men.

Maui participants had experienced:

- Drought, 10 (71%)
- Flooding, 6 (43%)
- Wildfire, 5 (36%)
- Hurricane, 4 (29%)
- Tsunami, 3 (21%)
- Earthquake, 1 (7%)

Experiences after the disaster (all participants):

- 7 (21%) *frequently* think or dream about their disaster experience
- 12 (36%) *frequently* discuss their disaster experience
- 21 (64%) are *very concerned* about future disasters
- 29 (88%) are now *more prepared* for future disasters

Preparedness for future disasters (all participants):

- 17 (52%) have a disaster kit at home
- 14 (42%) have some resources prepared, but do not have a disaster kit at home
- 25 (76%) have an emergency plan at home
- 24 (73%) have an emergency plan at work, but only 17 (52%) *practice* their emergency plan at work

Disaster stories from Maui: Drought

Drought is often considered a slow disaster, because it can last for many months or years. Maui has been experiencing drought for number of years, and the interview participants discussed a number of factors that are associated with drought. For one, because the drought has lasted for so long, many people who do not work directly with the land do not think about the drought or work to conserve water.

Luke said, "The fact that we have been having this endless drought for 10 to 12 years, you know, people – it's very easy to adjust along with the climate so we are adapting it unconsciously."

People who work with the land do not have that luxury. Many farmers and ranchers in Maui have experienced extreme losses from the reduction in rainfall. Additionally, natural resource managers must keep a careful eye on the health of the island ecosystems. They are always aware of direct and indirect issues, or the compounding impacts of the drought conditions. One example is wildfire. When brush and other foliage die and dry out during an extended drought, conditions become perfect for fires to spread.

Talia said, "The [wildfires] that impacted me most were the ones that shut down the highway... when they started letting people through, the hillside was still on fire... and you are there, stuck in bumper to bumper traffic, nowhere to go. Everything is burning above. There is stuff on the road and everything was kind of crazy, so it's those ones that cut me off from getting home."

Another example is the air quality issues and resulting respiratory problems in the community. These are slow developing and long-term impacts of this slow disaster.

Iris said, "The fugitive dust situation, that's incredibly bad... Fugitive dust, which is just the dust that comes from [the large agricultural fields] when they harvest... because we don't have the rain to really keep any of the moisture in... They create erosion so that when the wind comes, the dust blows, severely affecting communities."

The community is effected by drought in many ways as the island and near-shore environment are impacted by drought and related wildfires, erosion, and infrastructure damage.

Iris said, "That certainly impacts people too, and especially Native people who have cultural practices. And the fish are not there like they used to be."

Disaster stories from Maui: Flooding in Kihei

"Talia", "Luke", and "Iris" all talked about their experiences with the flooding that happened in Kihei in 2011, and the many compounding hazards from floods. Because of the ongoing drought conditions, the gulches around the island filled with soil, dried vegetation, trash, and other debris for a long period of time. When a heavy rainstorm hit the island, flash floods erupted. Kihei was hit with major flooding, as water filled the streets. The gulches had become full over the years, so mud and debris also washed down to the coast, and covered the town.

Iris said, "There was so much debris that was trash; refrigerators, things. People were throwing things over the gulches, or just lawn cuttings and trees. The velocity of the flooding as it was moving down the gulches and then it would get dammed up… It was shocking."

Luke said, "The first thing that came down was brush and other stuff that was just lying in the gulch… the retention basin fills up with brush and dead animals. When the water comes down, it lifts up the brush, knocks down my fence, pushes all the brush out in the ocean… and then the mud comes down…"

Talia said, "The flooding was so bad here. Everywhere, driveways got covered in silts, certain sections of the road were undriveable... It shuts everything down and people get out and help dig... There was a lot of damage to rural areas. Plus you couldn't use the ocean for weeks."

Kihei was covered in water and mud and debris. The community had to be proactive in their response because, as Talia said, "they had to get into their driveways." But it wasn't just the land that got covered. The mud and debris washed into the ocean, covering the reefs and damaging the near-shore environments. The marine ecosystem was harmed, and the beaches were out of bounds to people for weeks afterward.

Another hazard in disaster events like this is the psychological response to stress people feel when they are unprepared for disasters. They might make poor or even dangerous decisions they would not make if they were calm.

Luke said, "The water was still going... I'm checking things out and a woman walks down the road towards me and says, 'I have to get home, my husband was on the other side [of the floodwaters] over there, I have to get home to my husband.' And I go, 'You can't cross. You'll die.' She says, 'No, I have to get home. Can I use your phone to call my husband? He could come and get me.' 'No, we don't want your husband to die either.' So that's when we see what happens with people's brains when it's not the way they expect it to be. It's very easy to melt down. That what happened to this woman, quite amazing."

Disaster preparedness can include preparing your home or work environment with a disaster kit and an emergency plan.

Luke said, "We have a go-bag and we can either go straight up the hill if we need to or we can drive away a few miles if it will be safer. It just depends on the circumstances."

But disaster preparedness can also happen at a community level. Talia explained that the community in Kihei "overall feels kind of disempowered because they don't control what happens." The decisions made by property owners and land-users at higher elevations also impact the community, as was clearly experienced during the 2011 flood.

Community education and engagement are therefore also needed to increase preparedness. When more people learn and about disaster preparedness, the better off everyone in the community will be.

Iris said, "People didn't really realize because we've had drought for so long… I think we need better community education to everybody because people don't really understand the issues, things like dumping in the gulches that could be stopped."

Talia said, "There are some active parts of the community that spend their free time organizing workshops."

Disaster stories from Maui: Tsunami Warning

Maui has been fortunate in recent years not to experience a devastating tsunami. Several smaller tsunamis with localized damage have occurred. One story from an interview participant explains the compounding hazards of tsunami. "Doug" was working on a sugar plantation when they received word that a tsunami was heading toward the island. While tsunamis are out in the ocean, it is impossible to predict how bad it will be once it hits land, so every tsunami warning must be taken seriously.

"And we decided that we were going to shut down everything... not knowing [how big it was] we just said everybody go home, do what you can."

As they were closing everything down, they realized that a lot people from the communities nearby – "all these people in the hotels and even some of the local people that lived in low lying areas" – hadn't received very good communication about where to evacuate to. The roads became very congested, and evacuation traffic to higher ground came to a stand-still.

"A lot of them are tourists; they had no idea where they were going and it was kind of a danger from that standpoint... We opened all the gates to the field roads on the plantation so if anybody drove up the road or any place they could get to on the plantation at a higher elevation and park their car in the road or whatever.

There was no plan for this make-shift evacuation route. Doug and the others at the plantation saw a need in the community to get to higher ground, and quickly! So they reacted to the situation to help get people out of the stuck traffic on the main roads and up to safety at higher ground. But, Doug laughed,

"Afterwards we found out some people got stuck... or they got up there and parked and couldn't find their way out. And we never thought of that either, because it was just the emergency, just give them some place to go."

It was funny in hindsight because the tsunami turned out to be very small and caused almost no damage. The only impact on the plantation was loss of money and worker hours due to the evacuation. If the tsunami had been big, the quick thinking at the plantation would have saved lives – being stuck or lost is far better than being washed away.

"Thinking back it would have been good for us to have some preparedness, to say 'We'll go up there and find them and help them get out afterwards.' But we didn't [have a plan ready for this] because it was something new that we hadn't experienced."

Preparing for disaster in Maui

Think about where you spend the most time – are you prepared if a disaster struck there? You may want to increase disaster preparedness at your home or office. Maybe you drive around the island a lot and would like to increase your disaster preparedness in your car. Spend a little time now considering ways to increase your preparedness so that if disaster ever strikes, you will be ready.

The American Red Cross suggests increasing your disaster preparedness by following three simple steps:

1) Get a kit:

There are multiple types of emergency kits, and you should consider whether having one or more emergency kids would make you more prepared to respond to disasters common to the parts of the island where you spend the most time.

- Basic first aid supplies
- Evacuation kit with clothing, medicine, etc.
- Shelter-in-place preparedness kit

Note: Many people have elements of a kit scattered throughout their homes, but do not have an actual kit.

2) Make a plan:

Meet with your family or coworkers to discuss what you will do in case of a disaster event. Consider different scenarios, e.g., Morning versus evening; At home versus at work/school; Hurricane versus earthquake. To create a good emergency plan, you should discuss:

- How will you respond to emergencies at different places you are likely to be?
- Who is responsible to do what?
- What will you do if you are separated?
- Where will you go if you must evacuate?

Important: Review the plan and *practice it*. Everyone will benefit from practice, but especially children (for a home plan) or new hires (for a work plan).

3) Be informed:

- What are the types of emergencies that might happen in your area (**each** area where you spend time)?
- Who are your local authorities are, how to notify them, and how they will notify you during an emergency? (e.g., Sirens? Advisory, watch, warning?)
- What are local evacuation routes?
- Can you take community trainings (CPR, AED, CERT)?

SHARE WHAT YOU LEARN WITH OTHERS
What **you know** could save someone else's life.

Creating an emergency preparedness kit

What should be included in a good kit?

In the immediate aftermath of a disaster, emergency crews may be unable to immediately reach you. It is a good idea to have an emergency preparedness kit available in your home, office, school, and/or vehicle.

Take some time with your family or coworkers to consider what you would want to have in a **quick-grab evacuation kit** if you must leave quickly for safety; or in a **shelter-in-place preparedness kit** if you are trapped or must stay where you are for an extended period for safety.

Health and Wellness Items:

- Water – one gallon per person, per day
- Food – nonperishable, easy-to-prepare, especially requiring no additional cooking or water
- Manual can opener, eating utensils
- First aid kit
- Medications (7-day supply), other medical supplies, and medical paperwork (e.g., medication list and pertinent medical information)
- Sanitation and personal hygiene items

Safety and Logistical Items:

- Flashlight
- Battery powered or hand-crank radio (NOAA Weather Radio, if possible)
- Extra batteries
- Multipurpose tools (e.g., Swiss army knife)
- Extra cash
- Emergency blanket
- Copies of personal documents (e.g., proof of address, deed/lease to home, passports, birth certificates, and insurance policies)
- Family and emergency contact information
- Map(s) of the area

Additional supplies you might want:

- Whistle
- Matches
- Rain gear
- Towels
- Work gloves
- Tools/supplies for securing your home
- Extra clothing, hat and sturdy shoes
- Plastic sheeting
- Duct tape
- Scissors
- Household liquid bleach
- Entertainment items
- Blankets or sleeping bags

The emergency preparedness kit information on these pages is sourced from the Center for Disease Control and Prevention (CDC), the American Red Cross, and Ready.Gov. Please see the Resources section at the end of this booklet to find more resources to help you prepare for disasters.

What are YOUR special needs for an emergency kit at your home or work?

- Glasses /contact lens cases and solution?
- Baby care supplies, i.e., diapers and formula?
- Hearing aids?
- Cell phone / chargers?
- ?
- ?

_____

_____

_____

_____

_____

_____

_____

_____

_____

_____

Disaster and Climate Change Resources

Psychological Recovery from Disasters:

To access support or further information about post-disaster psychological recovery, please visit the American Psychological Association (APA) Psychology Help Center webpage on recovering emotionally from disaster at http://www.apa.org/helpcenter/recovering-disasters.aspx.

Psychology of Climate Change:

To read more about the psychology of climate change, please visit the APA Psychology and Global Climate Change task force website at: http://www.apa.org/science/about/publications/climate-change.aspx.

Disaster Preparedness Information:

Extensive disaster preparedness information is available from

- the Center for Disease Control and Prevention (CDC), http://www.bt.cdc.gov/preparedness/kit/disasters,
- the American Red Cross, http://www.redcross.org/prepare/location/home-family/get-kit, and
- Ready.Gov, www.ready.gov.

Climate Change Information:

The Pacific Regional Integrated Sciences and Assessments program aims to help Pacific Islanders prepare for and manage the risks from climate variability and change. The Pacific RISA is funded by the National Oceanic and Atmospheric Administration (NOAA). Please visit the Pacific RISA at www.PacificRISA.org.

www.ingramcontent.com/pod-product-compliance
Lightning Source LLC
Chambersburg PA
CBHW052047190326
41520CB00003BA/218